RATS
KW-128

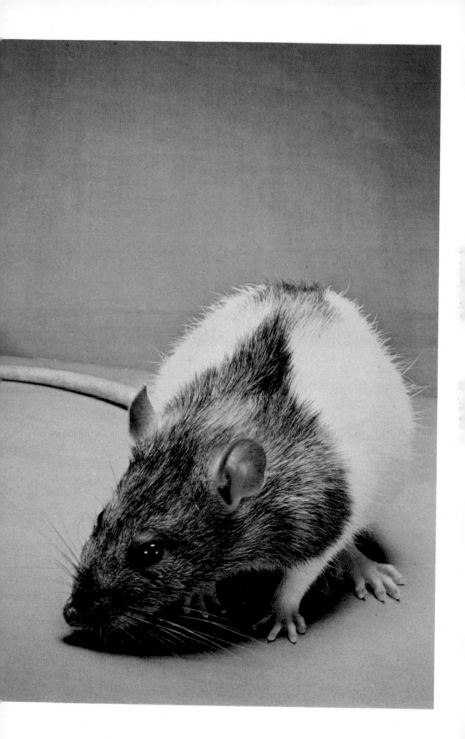

Contents

Introduction, 6
About Rats, 8
Why a Rat?, 22
A Home for Your Rat, 33
Food For Your Rat, 43
Taming and Training Your Rat, 55
Breeding Your Rat, 63
A Healthy Rat, 80
Index, 92

© **1988 by T.F.H. Publications, Inc.**

Distributed in the UNITED STATES by T.F.H. Publications, Inc., One T.F.H. Plaza, Neptune City, NJ 07753; in CANADA to the Pet Trade by H & L Pet Supplies Inc., 27 Kingston Crescent, Kitchener, Ontario N2B 2T6; Rolf C. Hagen Ltd., 3225 Sartelon Street, Montreal 382 Quebec; in CANADA to the Book Trade by Macmillan of Canada (A Division of Canada Publishing Corporation), 164 Commander Boulevard, Agincourt, Ontario M1S 3C7; in ENGLAND by T.F.H. Publications Limited, Cliveden House/Priors Way/Bray, Maidenhead, Berkshire SL6 2HP, England; in AUSTRALIA AND THE SOUTH PACIFIC by T.F.H. (Australia) Pty. Ltd., Box 149, Brookvale 2100 N.S.W., Australia; in NEW ZEALAND by Ross Haines & Son, Ltd., 18 Monmouth Street, Grey Lynn, Auckland 2, New Zealand; in SINGAPORE AND MALAYSIA by MPH Distributors (S) Pte., Ltd., 601 Sims Drive, #03/07/21, Singapore 1438; in the PHILIPPINES by Bio-Research, 5 Lippay Street, San Lorenzo Village, Makati Rizal; in SOUTH AFRICA by Multipet Pty. Ltd., 30 Turners Avenue, Durban 4001. Published by T.F.H. Publications, Inc. Manufactured in the United States of America by T.F.H. Publications, Inc.

RATS

BY SUSAN FOX

Rats are fun and good pets for people of any age. A black and white hooded rat is fun for this young man (*left*) and a self-colored rat for this young lady (*below*).

Introduction

This is a book about pet rats. As you read this book, you will notice that all rats and rat owners are referred to as "he" (except in the chapter on breeding your rat, which refers to female rats). This is not to suggest that girls should not own rats; at least half of the people who own rats are girls.

Pet rat care is described clearly and precisely. In addition to chapters on basic care there are other chapters (*e.g.,* mazes, taming and training) which tell how you and your rat can have fun doing things together. Rats are active animals. That is, they can be taught things, taken places and, most important of all, be your friend.

You can learn most about your rat through your experience with him and those of others. The information in this book is meant to help you enjoy your years with your pet.

About Rats

At some point in history the rat found advantages to living with man. He shared man's home and food and stowed away in his vehicles of transportation. The rat was resourceful but cautious in his dealings with man, which is why you still find rats sharing man's world. Competition for food between rat and man is intensified because the population of each is about the same. Though a rat cannot eat as much as a man, each day he can destroy a man's daily allotment of food.

KINDS OF RATS
There are hundreds of different species of rats in

Grooming is an inherent behavior of rats; both domestic and wild rats groom themselves, each other, and their young.

the world. Usually when one thinks of a rat, the picture of a dirty sewer rat comes to mind. Not all rats look like that; in fact, they are in the minority as far as species are concerned. Some rats have pretty squirrel-like faces with soft fur, large brown eyes and hairy tails, such as the **bushy-tailed woodrat** of the western United States and Canada.

The bushy-tailed woodrat is also known as the **packrat** because of his habit of packing off objects that catch his fancy. Silverware, shiny stones, nails, etc., can all be found in a packrat nest. The nickname "trader rat" originated because the rat would be seen carting something to his nest; when he would see something else that he preferred, he'd drop the first object and cart off the new one.

Another interesting rat is the **kangaroo rat,** which has long hind legs that carry it in bounds of six to

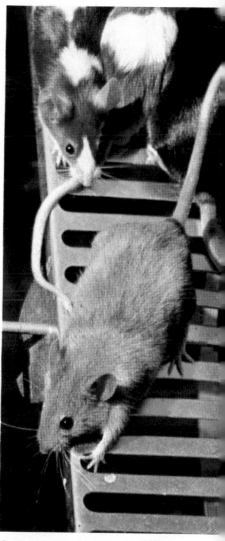

Caged rats need exercise for fitness. Provide your pets with a ramp or a ladder they can climb up or go down.

(*Left*) This 3-striped rat from West Africa is quite small, as can be seen from a comparison of its size to the grasshopper. (*Below*) However, the common rat (*Rattus norvegicus*) can grow to a foot in length, excluding the tail.

eight feet. When frightened the rat can jump 24 feet. This is truly remarkable for the small 15-inch rat. The rat has an eight-inch–long tail which it uses to change direction in midair. Its most notable characteristic is that it does not need to drink. It can manufacture its own supply of water from the starches in the seeds it eats. Kangaroo rats live only in North America.

The **African giant rat** is the largest rat. It measures between two and three feet in length including the tail. It lives in holes in the tropical forest.

Spiny rats are called

that because their hair has been modified into short, sharp, spiny bristles which discourage predators. They live in forests in South and Central America, usually near water, and they seek shelter near boulders, stumps or roots.

The infamous **black rat** and **brown rat** are the two rats which have plagued man throughout the centuries. They live in all parts of the world, although they originally came from Asia. They are the most serious animal threat to man since they carry the disease organisms of bubonic plague, typhus and food poisoning. These rats were

responsible for the death of the 25 million people in 14th century Europe who died from bubonic plague. The rats harbored the fleas which carried the disease-causing organism for the plague.

The black rat measures 14 to 16 inches including the tail. The tail is longer than the body and is covered with scales. It has large ears, a pointed snout

Rats in the wild inhabit burrows for protection against predators. This rat finds security in a loaf of bread.

and soft fur that is black, brown or gray. The rats are also known as roof rats and ship rats. The black rat was the pioneer rat of the westward rat movement, but the more aggressive brown rat ousted the milder black rat.

The **brown rat** measures 14 to 18 inches including the tail, which is shorter than the rest of the body. It has small ears, a blunt snout and coarse hair. It is believed that the brown rat stowed away on a timber ship leaving Norway, hence the name, Norway rat. Other names for it are barn rat, gray rat house rat and sewer rat. The white laboratory rat is the descendant of the brown rat and black rat albino strains.

RAT SURVIVAL / RAT SKILLS
Rats live in all parts of the world except on the highest mountains and the frozen lands near the north and south Poles. The brown rat and the black rat

Rats are curious; they have very a good sense of smell, can hear well, and can see fairly well.

journeyed from Asia and reachod Europo by ship or over land. From western Europe they spread to North and South America by ship. (Large ships are now equipped with rat guards that stop the rats from running up the ropes that moor the ships to the docks.) The brown rat reached America around the time of the signing of the Declaration of Independence.

Both the black rat and brown rat live in small areas. Usually these areas each measure 150 square feet. If there is a serious food shortage the rats will travel great distances in search of it.

Man has been changing the earth at a rapid pace and on a scale never before seen. Those animals that can't cope die off, yet there is little evidence that the rat seems so inclined.

Rats are agile climbers, looking like small tightrope walkers. Their scaly tails

1

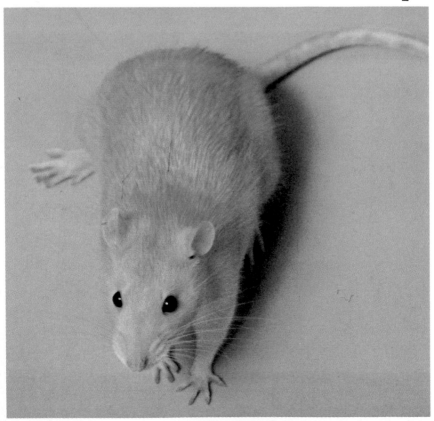

2

(1) Besides having cheek pouches, the African giant pouched rat is also much larger than the common rat. (2) A cream-colored fancy rat. (3) Drab coloration is of great advantage to the survival of a wild rat. (4) Ideally, the spinal stripe of this black hooded rat should be clean-cut and straight.

3

4

help them climb. The scales ruffle back and give the rat a good grip. Rats can also jump up to two feet. They are excellent swimmers and can stay submerged under water for three minutes.

The rat is intelligent. Some scientists believe rats are smarter than horses, cows and other animals. Rats are good learners. They learn from their own experiences and those of other rats; for example, if they see a dead rat in a trap they will avoid all traps.

Rats are highly prolific and can produce up to 100 offspring in a year. Few rats in the wild live more than a year because predators such as coyotes, dogs, cats, birds of prey, etc. are always on the lookout for a good meal.

RATS AND MAN
The rat has found his way into man's social institutions. The expressions "caught like a rat in a trap," "you dirty rat" and "cornered rat" are probably familiar to you. In the story "Cinderella," a fat rat is turned into a jolly coachman by Cinderella's godmother. The story of the Pied Piper is another one in which rats played a part. The piper played a tune on his pipe and lured all the town's rats into the nearby river where they drowned.

Ancient Rome considered the rat a good luck symbol. The Japanese had a traditional belief that the rat was a messenger of the gods. He is the first animal of their oriental zodiac which runs in twelve-year cycles. It is considered unlucky to be born in the year of the rat. In China the rat is a symbol of prosperity. In the Middle Ages in Europe the rats were believed to be creatures of the devil.

One can detect a hint of jealousy toward rats which seemed to have it so good. The rat was probably considered a good luck symbol and a sign of prosperity because of his easy access to the good life. He let man sweat and toil while he helped himself to the profits.

Rats in the wild are one of the worst animal pests. Besides spreading diseases they are responsible for damage totaling hundreds of millions of dollars each year in the United States alone. They destroy eggs, stored grains, fruits and vegetables. By taking a nibble here and there, and then soiling it, they make more food unfit for human consumption than they eat.

Rats also gnaw on furniture, lead pipes and the insulation on wires (sometimes causing fires). There have also been reports of rats attacking people.

Man has tried to exterminate rats by fire, floods, trapping and shooting. Some of the newest and more sophisticated methods of rat extermination are to treat food with anti-fertility drugs which do not kill the rat but make him unable to breed. Another new poison is the anticoagulant which

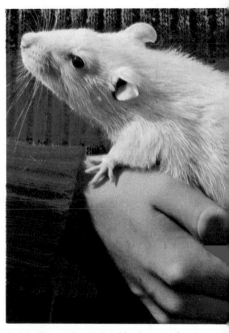

A healthy rat is alert, with clean and dense fur, ears free from scales or sores.

is put into food. The drug destroys the clotting factor in the rat's blood and the rat bleeds to death. (This drug is also used to destroy blood clots in humans.)

(1) Rats enjoy being petted and scratched behind the ears. (2) Feed your rat the same amount of food at the same time each day. (3) Never bring a strange rat close up to your face. Animal bites on the face are potentially dangerous.

18

2

3

Although rats have caused many problems for man, they are an important and necessary part of the ecosystem. They are so numerous that they form many important links in the food webs of many communities. If all the rats were exterminated from earth, some drastic things would occur. With one of their natural predators gone, the insects which rats eat would increase in population. The plant cover of the earth might look very different without rats scurrying around eating seeds. Carnivores would suffer and would most likely attack man's domestic animals for food in order to survive.

Man may distinguish between helpful rats and those that plague him in his cities and fields, but in general, rats are necessary to keep the world in balance. Large populations in man's cities are not needed, but the rat population must live on.

Rats are also important to man because of their usefulness in science. The white rat is used in many research projects and scientific experiments. Diet, drug effects, heredity, nutrition, learning and other aspects of human behavior are all projects which use rats. Many food additives, such as preservatives, food colorings and artificial sweeteners, are tested on rats. New medicines are first used on rats (and other laboratory animals such as mice, rabbits and guinea pigs) before they are released for sale to the public.

The rat has a unique history and has prevailed against extreme odds and has come out as a hardy, resourceful little animal. The rat's skillful abilities become apparent as a pet. He is a never-ending source of pleasure and can be a true friend.

PET RAT FACTS
Order:Rodentia
Family:Muridae

A rusty-nosed rat is a climbing species of rat found in Africa. It is a non-aggressive animal, but too high-strung to be kept in captivity.

Genus:*Rattus*
Species:*norvegicus*
Maximum life-span:2000 days
Average life-span:1000 days
Length:Head and body—80-300 mm
Length:Tail—80-200 mm
Weight:Adult—250-400 grams, Weanling—40-50 grams, Birth—5-6 grams
Heart rate:260-600 beats per minute
Blood pressure:115/90

Respiration rate:65-115 breaths per minute
Temperature, rectal:38.2⁰C.
Oxygen consumption:0.88 grams per hour
Age at puberty:45-75 days
Breeding season:any time
Gestation period:21-30 days (normally 21)
Litter size:2-24 (normally 8-12)
Sex ratio:46% female, 52% male
Weaning age:25-30 days
Female breeding life ends: 15-18 months

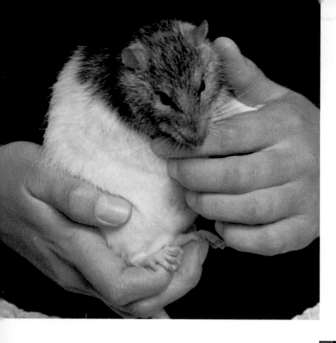

(*Left*) Hold your rat with two hands and snuggled against your body for greater security. (*Below*) Rats will eat their favorite food from your hand.

Why A Rat?

The history of its relationship to man has turned the general tide of feeling against the rat. Yet a rat makes a marvelous pet. Once you and your rat become friends he will come waddling up to his cage door and greet you with a lazy yawn, asking to be taken out. He can be tucked away in your sleeve or ride on your shoulder.

You do not need to bribe your rat with a piece of food; he will come to you willingly. Your rat can become as attached to you as a dog is to its master.

Rats are easy to care for with simple requirements of clean bedding and fresh food and water. They are inexpensive and odorless (if their cage is kept clean).

To an untrained individual mice and rats may look alike, especially during the early stages of development. However, an experienced handler can show you the features of a rat and a mouse.

When obtained from a reliable source, such as a pet shop, domesticated rats are as clean as any other domesticated animal. You will see your rat groom himself many times during the day and soil only certain areas of his cage.

The rat is inquisitive and intelligent. For example, he can be taught to come to you when you call him or how to climb a rope to get some cheese. He is a good pet for almost anyone of any age. Rats make good classroom pets. They can be handled and used in classroom projects. Unlike hamsters, rats rarely bite.

Rats are also a challenge to those who wish to breed them. Very little work has been done with them to develop different colors and hair textures. Hamsters, mice, guinea pigs and rabbits have been bred in many diverse colors with different textures to their coats, *e.g.;* long-haired, curly-coated and so forth. The field of breeding rats for these traits is still open and waiting for someone to pioneer. Perhaps that could be you.

CHOOSING YOUR RAT

It is best to get your pet rat from a pet shop or a friend who breeds rats. Be sure the cages in the shop are clean and not overcrowded. Animals coming from a crowded, dirty environment are less likely to make good pets.

The single most important thing to consider when choosing your rat is its health. Look for a lively, alert animal. It should be plump and have clear eyes. Avoid scrawny rats, those that act frightened and any with cuts or scratches, runny eyes or nose, wheezing and/or spastic movements.

Standing on two legs is a common stance of many rodents, including rats. Note how the tail supports the body, functioning like a third leg.

1 (1) A rat's amusing antics can capture your attention. (2) The albino rat, like the albino mouse, is indispensable in medical and genetic research. (3) Symmetry in the body pattern of body markings is difficult to achieve but possible through a long process of selective breeding.

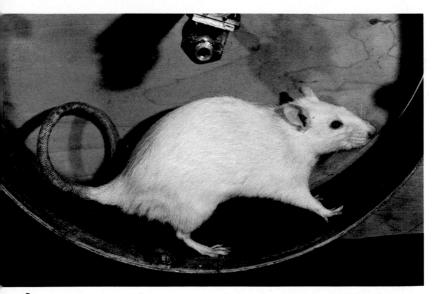

2

3

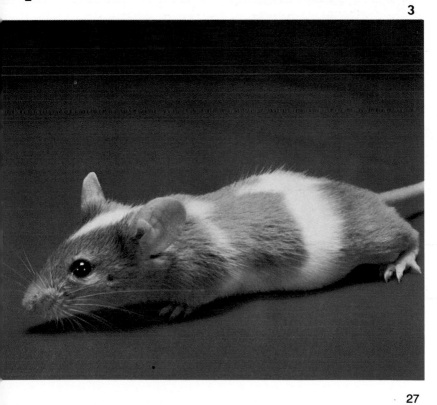

Why a Rat?

Two color phases are seen in rats: solid and hooded. A hooded rat is recognized by a pattern of color which covers its head giving a hooded appearance, The same color often runs in a stripe on the rat's back. Most rats are either black, cream, albino or agouti (each individual hair is banded with different colors). Some black rats are jet-black while others are a soft, dusty gray. There are various shades of cream-colored rats ranging from a light tan, which is scarcely distinguishable from white, to a warm golden color. The albino rat is white and has red eyes and pink feet, tail and ears. It lacks any pigmentation. The agouti most closely resembles a wild rat, so many people prefer it only in the hooded pattern.

Buy your rat when it is four to eight weeks old (about four to eight inches long). A young rat is more

The hooded rat is available in a variety of colors, such as black, brown, red, cream, gray, and possibly others.

You can play with your rat, tickling it with a feather. But, be sure not to get too close to the eyes.

easily tamed than an older one.

A typical rat has a long pointed snout, round peach-fuzz covered ears, a plump body and a long tapering tail. Some may have blunter faces than others, or shorter tails. Female rats are almost always smaller than males.

Should you buy a male or female rat? If you want a family of rats then you should buy a female. The female will, of course, need a male with which to breed, and in some cases a pet shop owner might let you breed your animal to one of their males. Most rats have a very sweet disposition, regardless of sex. If you get a rat that is "mean," get rid of it; this is an incurable trait.

COMPANIONSHIP
The decision to buy one rat or two is a matter of

(*Left*) The wire on the top of this cage separates two rats that would otherwise fight at every opportunity. (*Below*) A 10-gallon aquarium makes a good home for one adult rat. It is easy to clean and offers a good view of your pet at all times.

personal choice since rats do not need to be kept in pairs. If you decide to buy more than one rat, they should be of the same sex unless you want a family of rats. If two or more rats are kept together, you will have to buy a large cage.

Buying a second rat as a companion to the first one does not always work out. The older rat may resent the newcomer's intrusion into what he considers his territory. It might "pick a fight" by stalking over to the new rat's area of the cage and urinating on it. By doing this it is claiming prior right to the cage.

Rats can express anger. An angry one is an interesting sight. Its body will become very stiff and every hair will stand on end, giving the rat the appearance of twice its normal size. The long tail will start wagging slowly back and forth, picking up speed until the whole tail is trembling.

Rat fights are not a common occurrence because most rats get along well with each other.

If a fight does occur and you are there to break it up, you must be careful that you are not bitten. Use a towel to separate the fighting rats. Place each rat in its own cage so it can tend to any wounds. You can apply an ointment such as neosporin to any bad cuts. The best preventive measures are

clear. Any rats who appear not to like each other should be kept apart.

In all community animal societies, there is a pecking order which establishes the dominant animals. You should be aware of this if you plan to house a number of rats in one large cage. If any one rat is picked on, you will know that it is probably at the end of the pecking order. If you remove this one, the other rats would have to find a new victim. The introduction of a new rat into the cage would upset the balance and a new order would have to be established.

Two rats can become the best of friends. I have watched one rat dreaming in sleep, paws twitching, while its friend happily bit the sleeper's toes whenever the paw moved. Rats are playful animals, and if you have two rats, you will probably see them play-fighting with each other.

Rats can become good friends with some dogs, cats and other animals. Sometimes a rat and rabbit will get along as best friends. Those dogs that were bred originally to be "ratters" are not the best kinds of dog for your rat to meet. Some of these breeds are the terriers and schnauzers.

Rats should not be kept together with hamsters, mice, gerbils, birds or any other small animals. They will fight and the rat will probably kill the other animal. Aquariums should be covered if your rat is free to explore them because it would like nothing better than a freshly caught fish.

A single rat is perhaps the luckiest, for there is no competition with any other rat for your attention. A single rat might also be more devoted to you since you are the only one with whom it can play.

There are many types of housing available for rats. Most pet shops offer a wide variety of cages from which you can choose. Rats are active animals that do best in roomy cages. The minimum size cage you should look at is 16"x10"x10" to allow sufficient room for the young rats. Be sure that the spaces between the wires are not large enough for any of them to fall through. Rats can gnaw through many materials; metal is one of the few that defies their teeth. Be wary of a cage that has large rectangular spaces in the wire mesh, for your plump rat will flatten himself out and slither through a space.

Aquariums offer one of the finer choices of housing. They come in many sizes, are easy to clean, can be attractive and make it convenient to watch your pet's antics. In addition there is no worry about shavings and food falling out and making a mess since the bottom portion is all enclosed. The simplest way to clean your

A simple all-wire cage with a metal or plastic bottom tray will be adequate for housing a single rat or a pair of them. The water and food dish are installed separately.

rat's aquarium home is to buy a metal scraper and scrape up the shavings. You can wash the glass sides to keep them sparkling clean. The top should be covered with a wire screen to prevent your rat from escaping. The wire screen can be held securely to the aquarium sides with clips purchased either at a pet shop or a hardware store. No matter how deep the aquarium, cover it. Rats can jump as high as three feet!

A large bird cage can make a satisfactory home for your pet, too. Your rat will enjoy making use of the perches, climbing from one to another. However, most bird cages have doors that were not designed to withstand a rat's curiosity—buy clips to prevent your rat from escaping.

Wooden cages are not as desirable as metal ones because rats can chew through wood. If you do use a wooden cage, the wood should be at least

A wire cage with solid walls at each end can protect your rats from exposure to draft. The wire handle also makes this cage easy to carry and move about.

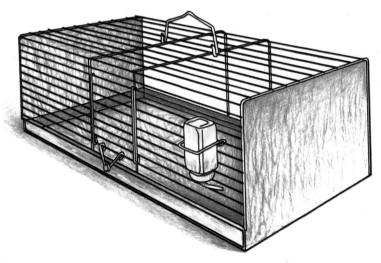

one fourth of an inch thick to discourage your rat from chewing his home into dust. You can reinforce the sides and edges with a fine wire mesh. The greatest disadvantage of wooden cages is that the wood absorbs urine and other liquids, and if the bottom is made of wood, then it will become smelly and unhygienic and eventually will rot.

Building and designing a cage for your pet can be fun and well worth the time if you remember a few simple facts. Don't build a cage with a wooden bottom. You can buy sheet metal at a hardware store and have it cut to fit the bottom of the cage. You can then make a frame out of wood and enclose all sides with wire screen or have wooden sides and a wire front, top and back. Be sure to cut a door large enough for your hand to fit through comfortably. File any sharp edges to prevent your hand being

Install a well-made drinking bottle that has a strong metal spout, not glass or plastic.

cut when you reach into the cage. Set aside an area within easy reach of the door where you can hook up the water bottle and food dishes. Use your imagination, but don't get carried away when you design and build the cage. You have to keep it clean and if it is full of nooks and crannies cleaning it will be a hard job. To ensure adequate ventilation, at least two sections (front and top, preferably) should be of wire screen.

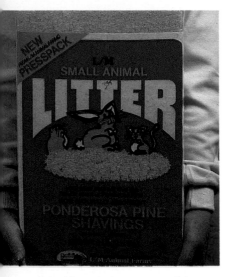

A bedding of whatever material you choose will keep the cage smelling sweet, provided it is replaced periodically when soiled.

BEDDING

The purpose of bedding is to keep your pet warm and clean and to keep the cage clean and smelling nice. Each kind of bedding varies in its ability to do the job. The ability of a bedding to absorb liquids and urine depends on how finely ground it is. The coarser ground shavings (called chips) do not absorb as well as do fluffy flakes.

Among the most widely available kinds of bedding are pine shavings. Though varying in how finely they are ground, pine shavings are good absorbers. They are cheap in nearly all regions of the country. However, pine shavings do not deodorize well.

Chlorophyll shavings are usually dyed green. Their most important characteristic is that they are excellent deodorizers. They do absorb if they are fine enough. Since they have been chemically treated, they are unfortunately more expensive than the pine.

Cedar shavings are another fine choice, as they have the same characteristics as the chlorophyll shavings. They, however, have a tendency to stain your animal's coat if your pet lies on them

The carrying carton from the pet shop is intended for temporary use only. A rat can chew its way out, if kept in it indefinitely.

when they are wet or dirty.

Another cut of wood is called feathers. They are soft, long and stringy. Feathers are absorbent and your rat will enjoy burrowing through them.

If you use sawdust, it should be made from softwoods. Sawdust rates about the same as pine shavings except that it has a higher absorption potential. A disadvantage which must be considered, however, is that sawdust is dusty and can penetrate your rat's fur or occasionally get in his eyes.

Straw is another option. It does a better job when used with shavings. Your rat will enjoy making nests in it. You can use the grass from your yard as long as no sprays or fertilizers were used before you cut it. Let it dry in the sun before putting it in the cage; if it is not dry, it will mildew. With straw and grass there is always the risk of introducing mites or other insects. Examine the straw carefully before using it in the cage. If an infestation does occur, dispose of the contaminated bedding, wash the cage and disinfect both it and the area around the cage with a flea spray. Because it is milder, use a cat flea spray rather than a dog flea spray on your rat.

Peat moss is an excellent deodorizer. Its acid content helps counteract the decomposition of animal droppings. It is expensive and hard to find in some areas, though.

If you have a large number of animals be sure to investigate buying a large bale of shavings at a feed store. This will save you time and money.

CAGE ACCESSORIES

An empty room looks uninviting to most people and in the same sense a cage which offers your rat no privacy may not make him feel at home. You can give your rat a nesting box made out of an empty (washed and dried) milk carton or an empty oatmeal container. This box could be compared to your own bedroom. You can give your rat an old sock or shirt to put in his box. Be sure to rinse it out each week so it will stay clean. Call your rat out of his box; don't reach in and grab him. This is his private box.

An exercise wheel for your rat is not as necessary as it is for hamsters. However, baby rats are quite fond of playing on the wheel, spinning each other round and round. It is not a good idea to get a wheel if your cage is too small. There is a multitude of things that you can buy or make for your rat to enjoy—ladders, mirrors, pine logs, etc.

You can buy a water bottle at a pet store. There are glass or plastic models available. Use a gravity or demand bottle to provide fresh clean water. If you use a pan or dish, the water will become a dirty mess because your rat will fill it with shavings, droppings and food.

A food dish is better

than nothing at all. It will prevent your rat's food from becoming contaminated with his droppings and bedding. Have a separate container for any moist foods you plan to feed your rat.

WHERE TO KEEP YOUR RAT'S HOME

Rats can be kept either indoors or outdoors as long as the temperature does not drop below 40 degrees. There should be no objection to keeping your rat in the house, since they are almost completely odorless, smelling mostly of urine if the cage has not been cleaned often enough. (Mice and hamsters have distinct, detectable odors which are more noticeable in the males.) If you keep your rat's living quarters clean then there should be little odor. Your rat's cage should be able to fit nicely into any room arrangement. You can keep the cage directly above your supplies,

A metal spout on a drinking bottle stands up to more wear (and potential abuse) than a plastic or glass spout.

resting it on a box or shelf.
Don't place it where it might get too hot (near a heater outlet) or too cold (near a window). Don't leave any clothing, medicine or detergents near your rat's cage because at night your pet's little paw will come sneaking out from between the wires, ready for mischief. A cage with wire

exposes your rat to drafts which can give it a cold. Whether you keep your rat inside or out, make sure that it is not exposed to any drafts.

If you keep your pet outdoors, the cage can be kept in a garden shed, patio or rabbit hutch arrangement. Your rat will grow a thicker coat against any cold weather. Do not place the cage where it will get direct sunlight or a cold draft. During the winter months bank the shavings up high in the cage and provide your pet with an extra old rag or two to help keep it warm. In the summer leave a thin layer of shavings on the cage floor so your rat can keep cool.

Special care must be taken that the cage remains clean so that no wild rats are attracted to it. It is up to you to protect your pet from cats. Be sure the cage door is always firmly shut. Some rats might not be afraid of cats so be careful to keep them

out of harm's way. Certain breeds of dogs such as schnauzers and Yorkshire terriers are "ratters." That is, they were bred to kill rats and they have an instinctive dislike of rats. If your rat is going to share your home with one of these dogs, you will have to take extra precautions to prevent your pet from coming into contact with it. This means you will have to close doors to the room where you keep your rat.

CLEANING YOUR RAT'S CAGE

How clean you keep your pet's cage plays an important role in its health and life-span. Most rats will urinate and defecate in one corner of their cage. If your rat has not yet established such an area, it probably will. This area can then be cleaned out

FACING PAGE: Fancy rats with light coloration (white, beige, etc.) need an extra clean cage to keep the fur in pristine condition. The bedding should be changed more often, too.

approximately every three days to minimize odors. If you clean it too often, your rat won't keep it "established." You can use chlorophyll or cedar shavings in this corner to help control odors.

Clean the cage at least once a week. Each cleaning should include scrubbing the water bottle, especially the tube where the water comes out, making sure there are no food or shavings clogging the tube. Scrub the food dish also. Be sure to give your pet a new bedroom frequently. Once a month give the cage a good scrubbing; disinfect both it and the area around the cage. Metal cages have a tendency for the wires to get grimy. If this is difficult to remove, you can take a metal file or some steel wool to help you rid the cage of it. Keep your eyes open for any sharp corners or rusty spots; cut or file them away.

It is not a good idea to place clean shavings over old because it gives a false impression of cleanliness when in truth you could have several layers of filth below. Scoop out some of the dirty shavings and then add clean ones.

Remember, next to a good diet and clean water, a clean cage is the most important thing for your rat's health. It is his home, so keep it clean.

A good time to get to know your pet is when you clean the cage. If you keep your pet on your shoulder, it can keep you company so cage-cleaning is more enjoyable for both of you. If you decide to let your rat have the run of the room when you clean the cage, be sure to close the door and make it "rat-safe;" that is, have nothing around that it could eat that could cause sickness if eaten. You can always put your rat in a bucket or a small carrying cage. These cages are a good idea to have around in case of an emergency.

Rats have lived on this planet for so many eons under various conditions because they eat almost any kind of food. Pet rats are easy to feed. Like man, rats are omnivorous, eating fruits and vegetables, grains, seeds and meats.

WATER

Water is a necessity for your rat. A rat drinks about an ounce of water daily. You can buy a gravity-demand water bottle at a pet shop or, if you prefer, valve-type water bottles are also available. The bottles come in different sizes, but one with an eight-ounce capacity is ideal. The bottles are made of either plastic or glass.

The gravity-feed water bottles operate only when the animal sucks or licks the opening. Then air replaces the water lost. These bottles can leak if food or shavings are caught in the tube. Sometimes the position in which the bottle is placed will make the water siphon out. There are various

If you ever permit your pet rat to stay out of the confines of a cage, be sure that all exits are secure. Once out of the house, retrieval may prove impossible.

Rodents have the ability to hold on to their food. This is an African grass rat enjoying its favorite food.

types of bottle holders. In a wire cage, a piece of metal is looped around the bottle. In an aquarium, a metal holder is hung from the top and keeps the bottle upright and inside the rat's home.

Do not give your rat water from an open container since he will purposely spill it or fill it with his bedding and, thus, his droppings. Your rat does not need to drink anything other than water. Occasional liquids will be fine as "taste treats," such as a sample of the soup you had for dinner, but his liquid requirement is water.

Change the water frequently, at least every three or four days. As part of your normal routine give the bottle a good scrubbing each time you change the water. Clean the tube where the water comes out and remove anything which might be blocking the flow.

FOOD
A complex diet of well-balanced, wholesome and fresh ingredients is necessary to provide adequate nutrition for your rat. Your rat's diet should be based on a protein source and a fruit and vegetable mix. Good protein sources are dry dog kibble, fish flake food, freeze-dried brine shrimp

Milk may be a good food in general, but it may cause stomach upsets in a rat.

(50% protein) or plankton (69% protein) and various grains.

A fruit and vegetable mixture should be considered essential for your rat because of the vitamins and minerals it can provide. Use a variety of fresh fruits and vegetables chopped into rat-size bites. To make such a mixture, simply add finely chopped fruits, *e.g.,* bananas, apples, strawberries, etc., to chopped lettuce, grated carrots, diced cucumber, tomatoes, etc. Make no more than five days' worth at a time. Store in an airtight container in your refrigerator. A separate dish should be used for any moist foods you feed your rat.

Uneaten moist foods should not be left overnight in your rat's cage. Wash and dry the dish to keep it clean. Start off with a small amount, about one teaspoon, because your rat must work up a tolerance to the greens so he won't get diarrhea. Eventually you will be feeding your rat from two to four teaspoons every night or every other night. All of the sample diets listed below include the fruit and vegetable mixture.

SAMPLE DIETS
1.) Breeders of large numbers of rats find it easiest to feed a high quality animal cube or pellet (usually called lab diet or small animal diet) than to spend time mixing various foods together. These pellets contain at least a minimum of a balanced supply of the basic essentials to maintain a healthy rat. Read the package analysis of different brands to see what kind gives you the most for your money. Prepackaged foods lose some of their food value if they have been sitting on the shelf too long. Read the manufacturing date and don't over-buy a supply. Do supplement

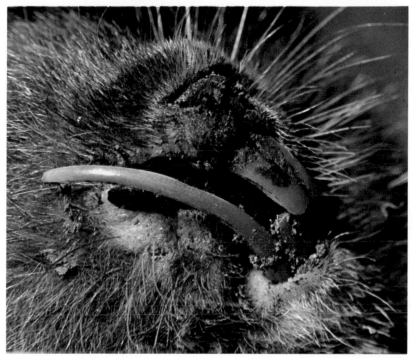

Your pet rat if deprived of normal chewing can develop a dental abnormality similar to the condition shown here. The lower incisor of this mole rat is now too long for efficent chewing of food.

with the fresh fruit and vegetable mixture.

2.) To supply adequate protein to the typical hamster mix, which contains grain, seeds, nuts and alfalfa pellets (which most rats don't eat), you can add dry dog or cat kibble. This will be a satisfactory diet for your rat. Add the fruit and vegetable mixture.

3.) Crackers, breads and cereals found in your kitchen cabinets are readily accepted by rats. You can add a hamster mix to the "cereals" and feed as a complete diet; don't forget the fresh fruit and vegetable mixture. Only

Rodents, not only rats and mice, can manipulate food dexterously. Their front paws have prehensile digits, except the thumb which is the least developed.

unsweetened foods or those with a minimum of sugar should be used. Rats can get addicted to sugar and it is not good for their teeth. The cereal mixture is a pleasure to give your rat. It will enjoy eating the different foods. Here your rat can choose what it wants.

Now that you know what to feed your rat, you need to know how much. Throughout the rat's life there are certain requirements regarding the quantity of foods to be fed. A rat's weight can fluctuate if it is fed sporadically. Feed your rat the same amount of food and preferably at the same time each day. Since your rat is nocturnal and busiest at night, you should feed it in the evening rather than in the morning. Give your rat approximately one-third of a cup of food each night. Rats store surplus foods. Food such as grains or dog kibble should always be available to your rat since it will nibble on this throughout the day.

While a rat is still forming as a fetus in the female rat, all its needs are taken care of by her. Faulty nutrition in the mother's diet will show up in the young. During pregnancy and lactation a female rat should be free-fed; that is, given all she can eat. A nursing mother can eat her own weight in food in two days, so give her at least two to three times as much as otherwise. You can find more about this in the discussion on breeding.

Once a young rat starts eating hard foods (usually at the age of two-and-one-half weeks) its appetite will

increase tremendously and it too should be free-fed. By the time your rat is four months old, the growth rate should have slowed and you will then want to start stabilizing the amount of food you give it. A

You can always lure your pet rat out of the cage by offering it some food.

mature rat should be fed the same amount of food on a daily basis to help maintain the naturally plump figure. If you think the animal has lost or gained a noticeable amount of weight, and you know this is not due to illness, adjust the amount of food accordingly.

Some rats lose weight when they get old. If you notice this, you must increase the amount of quality foods that he likes and cater to his wishes so he can regain his weight.

Whatever you feed your rat, always use a dish for your rat's food to keep it from coming into contact with his bedding and droppings. His food supply should be kept in an airtight container, such as a coffee can or glass jar with a lid, to keep it fresh and to prevent spoilage. Store the food in a location convenient to your rat's cage. If the cage is on a box you can use the box as a "storeroom" for your pet's supplies, including its food. When mixing various dry foods together you can put them into a clean, dry container (such as the empty coffee can) and shake it until the ingredients are mixed, or you can stir them together with a spoon.

These long-tailed thicket rats from Africa are enjoying fruit in a garden hedge. They are also known to feed on insects.

SNACKS

Rats relish table scraps, from cornbread to steak. When feeding table scraps use a container for moist foods or anything that is still too warm, remember to remove uneaten foods. You can give your rat rice, soups, potatoes, small pieces of meat (although your rat really does not need to eat meats) or a portion of a sandwich.

Both domesticated and wild rats enjoy eating insects. Mealworms can be bought at pet shops and are a nice treat for your rat. If you do make it a habit to feed your rat insects, be careful that you get them from areas that have not been treated with insecticides. Most rats will readily eat grasshoppers, katydids, moths and caterpillars, although you will find that some rats are afraid of the insects.

Your rat's teeth are constantly growing, but this growth is at a very slow rate and cannot be noticed by you. If they did

A sampling of food that rats and mice enjoy: oats, sunflower seeds, and dog biscuit. Grain is a natural rodent food.

not grow they would soon be worn down by all the hard foods he eats. You can provide your rat with hard foods to gnaw on, nuts in shells and rawhide dog bones as possible choices. If hard foods such as dog kibble are provided in the regular diet you need not worry about long teeth. Some rats love to gnaw, while others never feel the urge to do so.

Many people like to provide their pet with treats for special occasions or just to say "I love you." It is lucky for both the rat and the owner that the market abounds with special goodies to feed your pet. You can feed treats that are for cats, dogs, hamsters, birds and just about any other animal.

VITAMINS

Rats are susceptible to vitamin deficiencies. You can buy a vitamin supplement for small animals at your pet shop. It usually comes in a liquid form. You can feed it directly to your rat by letting it lick it off the dropper or you can add it to the drinking water. Follow the directions on the label.

The vitamins listed in the chart below are those that can be lacking in a rat's diet. The foods which contain the vitamins are listed in forms that are easily available to your rat. For example, many vitamins can be found in eggs, but eggs can be messy so you can feed egg biscuits of the type manufactured for birds.

The deficiencies listed are those that you might observe in your rat. If you read what a vitamin does and what happens when it is missing, you will understand the symptoms of the deficiencies. For example, if your rat has a deficiency of folic acid he can have nutritional macrocytic anemia. Folic acid is necessary for the formation of red blood cells. Immature red blood cells cannot perform important functions, such as transporting oxygen, food and waste. Thus, a rat is weak and emaciated.

Some vitamins are not listed because almost any diet that you feed your rat will include them. For example, thiamine, or vitamin B_1, is in grains and most rats usually get grains. Your rat gets the supply of vitamin D by licking its fur. A vitamin D deficiency can lead to rickets, but this is rare because many foods are fortified with the vitamin.

Usually there is not only one symptom of a vitamin deficiency because various nutrients intereact. Effective vitamin therapy is a diet rich with variety.

VITAMIN A

*Sources: Egg biscuits, (sold for birds); fish flake food (with a high content of fish products); green and yellow vegetables

Effects: normal function of skin cells and vision

**Symptoms of Deficiencies:* night blindness, lowered resistance to infection, conjunctivitis

Note: Rats use large amounts of vitamin A which they store in the liver.

VITAMIN C

Sources: citrus fruits, tomatoes, green peppers, cabbage, potatoes

Effects: wound healing, formation and maintenance of intracellular material in bones and soft tissues

Symptoms of Deficiencies: scurvy: swollen bleeding gums, loosening of teeth, weak bones and hemorrhages throughout body (sores or scabs may be evident)

FOLIC ACID

Sources: green leafy vegetables

Effects: maturation of red blood cells

Symptoms of Deficiencies: nutritional macrocytic anemia— immature red blood cell release—weak emaciated animal may be evident

PANTOTHENIC ACID

Sources: egg biscuits, vegetables

Effects: fat, protein and carbohydrate metabolism

Symptoms of Deficiencies: impaired coordination, easily fatigued

VITAMIN E

Sources: fortified foods, wheat germ oil, grains, green foods

Effects: essential for normal reproduction in both sexes

Symptoms of Deficiencies: males—sterility, females—death and reabsorption of fetuses

Note: Rats have a good store of vitamin E. It is thought that the mother gives her young a sufficient supply to last for the first one or two breedings. After that period a deficiency may show.

* Recommended feeding sources (ease and availability)

** Deficiency symptoms most evident in rats

The white rats used in the laboratory make good pets; they are docile, intelligent, and affectionate. They are also free from genetic defects that are found in highly inbred strains of fancy rats.

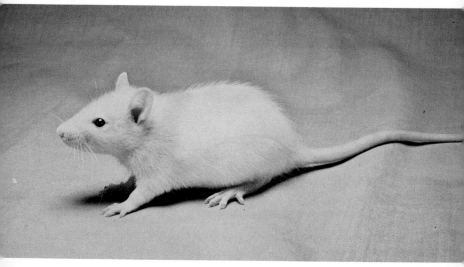

Before you can begin either to tame or train your rat you should decide upon a name for it. Your rat will then come to associate hearing that name with feeding time, a scratch behind his ears, exploration and any other method of attention that you may use. Thinking of a name for your pet can be fun, and you might want to enlist suggestions from others and then decide on the best one.

Once you have decided upon a name you will need to think of another way to call your pet to you. A clicking noise made with your tongue, "tch-tch," or a soft whistle are examples. A vibration also aids in getting its attention. You can tap lightly on the floor with your fingers or lightly clap your hands. Don't do it too loudly or you might frighten your pet.

HANDLING

When you are holding your rat, you do not have to squeeze it. Hold it gently with two hands and snuggle your pet up against your body. One hand should support the front paws and shoulder area and the other one should cradle the hind legs.

The instinct to flee is strong in a rat. A pet rat inevitably will try to escape when restrained for the first time.

Young rats can be taught to stay on your shoulder as a means of carrying them around. (This is more difficult to teach to older rats; they seem to prefer the security of your hands.) Rats can balance on your shoulder quite well. You should not have to worry about your pet falling off.

Wear clothing that will protect your shoulders from your rat's nails, which are tiny but sharp. When your rat is first learning how to stay on your shoulder, it might climb down and you could get scratched. If your rat climbs down, replace it and keep your hand ready to steady it. It will not take your pet long to learn this if you are patient and persistent.

Don't pick up your rat by the tail. It is an organ of balance and can be damaged by rough handling. When picking up your rat, scoop your hand under the body and lift the animal into a holding position. If you need to handle a strange rat or one that does not look friendly, it might be safest to pick it up by the scruff of his neck.

TAMING

Taming your rat means that you must gain the animal's trust. Fortunately this is easy and enjoyable to do. It is a rewarding experience to watch the shy rat that you brought home on the first day grow into a confident friendly rat. A rat that is handled frequently makes a better pet than one which is left by itself for long periods of time. Play with your rat often, at least four or five times each week.

You can start taming your rat by holding it in your hands and letting it get used to you. Remain near the cage and let your pet jump back into it when it wants to do so. Curiosity will prompt it to climb back out of the cage to see you. You can feed tidbits of food to increase trust. If

By nature an active animal, a pet tame rat will stay confined in a crystal bowl only for a short while. It will soon try to climb down and move elsewhere.

you take your rat out each day, it should be tame in one week's time.

Once your rat knows you, you can start to build up basic taming. If you expose your pet rat to many new and different environments at a young age, it will not be as frightened of strange environments when it gets older. A new environment can be a different room in your house, outside on the lawn or a visit with the family dog. Always allow your rat a few days' time to get used to any new surroundings. Because you are familiar to your pet, when introduced to new surroundings it will begin all its explorations from you. When it is frightened it will come running back to you. When you allow your rat to explore, be alert that your pet does not escape.

TRAINING

Once your rat is trained it will come to you when it is exploring. You should be very certain your rat knows you before allowing it to explore. While your pet is roaming around you can begin training it to come to you when called. This is most successful if the rat is in a room that he already knows.

Call by name, "tch-tch" (or whatever you decided upon) and a vibration to attract its attention. Coax your pet to you the first few times by talking softly. When it comes to you praise it. Pet it or offer its favorite food. Your rat will soon get the idea and will enjoy coming when called.

There are many other things which you can teach your rat to do. It can be taught to go from certain rooms in your house into the one in which the cage is kept. Set it down on the starting point and guide your pet in the direction you want it to go. When the destination is reached, give praise or a piece of food. Your rat will catch on to this one fast because it would much rather be in the room with its cage.

You can attempt to teach your rat the command "no." However, this is difficult because you can only use it when your rat is doing something you don't want it to do. The best method is to say the word "no." If this does not work, clap your hands once. You are startling your rat so that it stops doing what you did not want it to do. Whether the animal will learn the command "no" is questionable.

Your rat can be taught to do all sorts of wonderful things. It will do what you ask for three main reasons: a reward of food, the enjoyment the rat might get or because of the special attention you give when it has behaved well.

If your pet rat performs for either of the last two reasons you have a true friend. A food-trained rat comes to you not necessarily because it likes you but because of the food which you offer. It is far more satisfying to use yourself as the bait. When your rat does perform for you, you know it is because it wants to please you and receive attention.

Knowing your rat's moods might help taming and training go more smoothly. Your rat displays moods with certain signals. For example, a happy contented rat will grind its teeth together to let you know that it is pleased. If you hold its head in your hand with your fingers under the chin and the palm of your hand over its head, your pet might start grinding its teeth. Some rats not only

grind their teeth but also lick your hand.

A nervous rat will rapidly groom itself. It will usually stay in one spot, lying low against the ground with a blank look on its face. Another sign is body elimination in the form of a "dropping." This is part of the body's "fight or flight" mechanism. You can learn more about this by reading a physiology book on the nervous system.

Rats do not have good vision. They can detect a motion several feet away but cannot focus well on the object causing it. Vibrations help a rat to know if someone or something is coming, and the rat's sensitivity to vibration can be used by you when training your rat (for example, by clapping your hands to indicate "no"). The whiskers which are found near the eyes, lips and cheeks greatly aid the rat in its wanderings. When the rat is running along, the whiskers extend out in front of the snout.

The ones on the cheeks feel for horizontal surfaces while the ones above the eyes detect surfaces above.

MAZES AND PUZZLES

Rats have been used in mazes for many years to study behavior and what affects it. Whether you have to construct a maze for a project or whether you want to do it for fun, it is an interesting and rewarding experience. You will be amazed at how quickly your rat will travel through the maze you have constructed once it learns the correct route.

A maze is a network of passageways; some lead to dead-ends and others lead eventually to an exit. There should be an entrance, an exit and paths that lead nowhere. It is up to you to design the maze, and the designing itself is a very enjoyable activity.

There are many different kinds of mazes to make. You can make yours from cardboard, masking tape,

string and paint. Making a maze does not have to be hard work or expensive. For a more permanent maze you might want to use wood. A wire screen or high sides should be used to prevent your rat from taking the easy way out.

The most common type of maze is the one with passageways. The rat is placed in the maze and allowed to find its own way out. Continuous repetition soon engraves the correct path into the rat's mind and eventually it will go speeding through without any wrong turns. Time the rat to see what a difference there is from the first try to, say, the tenth try. Another kind of maze teaches the rat to read symbols. The example used is the odd/even concept. Four pieces of cardboard are painted with black stripes. Two of them have horizontal stripes and the other two have vertical stripes. Three of these (two horizontal and one

Newborn rats seek warmth and protection by creeping underneath their mother or staying underneath the bedding.

vertical, for example) are tied into a box so that they swing like doors.

Place the rat into the box. The only way for it to get out is for the animal to go through the door with the vertical stripes which swings open when pressed. The other two doors are blocked so they cannot open. Once out the door the rat receives a piece of food as a reward. After the rat learns which door to use for the exit, change the order of the doors, *i.e.,* move the door with the vertical stripes to a different position with the horizontal doors. Then you can change the doors to include two vertical doors with only one horizontal door, and eventually change their order. Soon the rat will grasp the concept of odd/even. You can use colors in the same way. Paint one door black, one gray and leave the other white. Have the black door "open." Again change the order of the doors until your rat recognizes the black door as the exit.

You can combine all that your rat has learned, odd/even and colors, into one puzzle. For example, paint the word "exit" on three doors. On one door, paint the word vertically in black letters. On the other two doors, paint the word "exit" horizontally; on one use brown letters and on the other outline the letters in black. The rat can learn to go through the door which has the word "exit" painted vertically in black letters.

You can also construct a "crazy" maze for your rat which can serve as a playground. Use a large box and set up ladders, logs, mirrors, bells and tunnels in an intricate manner. Be sure everything is fastened securely. Your rat will enjoy playing in it.

A study performed by a group of scientists found that animals exposed to a variety of stimuli (ladders, bells, mirrors, swings, tunnels, etc.) arranged in an intricate manner at a young age develop a larger cerebrum than animals that just sit in corners all day. This is noteworthy because the cerebrum is the area where thought processes take place. One could conclude that if there is a larger cerebrum there is increased capacity for learning. Perhaps you could do an experiment to find out if rats exposed to such stimuli do make better learners than those who just sit around.

Let your imagination think of new ideas. This chapter tells of only a few of the many possibilities.

The first thing to consider before breeding your rat is finding homes for the young rats once they are weaned. This is not difficult, but it is something many amateur breeders often forget about. Friends, teachers (who can keep them as classroom pets) and pet shops are all possibilities. You should be aware that most rats sold in pet shops end up as feeders (snake food). If you wish to avoid this you can hand-train your young rats before selling them to the pet shops and request that they be sold as pets if possible. Most laboratories breed special strains of rats to use in their research and will not be interested in your rats.

If you are going to breed rats for profit you will need large numbers of quality animals that are fertile, producing at least eight to twelve young rats in each litter. You must consider the fertility of both the female and male rats. In some cases, the male may not be a good breeder. You must also realize that taking proper care of all the rats will require a great deal of time. Good sanitation and clean surroundings are especially important if you want to breed healthy rats.

Breeding rats is an interesting experience whether it is for profit or pleasure. It is a joy to

A view of the cages for rats in the breeding room of professional fancy rat breeder in England.

watch the young rats grow from small pink creatures into furry bouncy rats.

The minimum breeding age to consider for breeding is ten to eleven weeks. Females should weigh at least 200 grams and males at least 300 grams. The best age at which to begin breeding is four or five months. The rats will be stronger and more mature.

MATING AND PREGNANCY

If the female rat is too skinny or too fat she probably won't become pregnant. During the heat cycle (which lasts twelve hours and usually begins in the late afternoon or evening) the female's genitalia become swollen, and if impregnation occurs the vaginal tract swells shut. After the male and female have mated (you might want to keep them together for a few days to ensure that the female is pregnant), remove the male and keep them in separate quarters. Some breeding systems do not require that you remove the male rat. An example is the harem system, in which one male is kept with a group of females, usually one male to every five females.

The cage of a pregnant rat should be at least 18″ x 10″ x 10″. The pregnant female needs plenty of room to organize her nest, food area and toilet. She will also need room to rest away from her young when they become demanding. Young rats require a large amount of room to run and play.

It is essential that the female rat be fed the very best foods you can give her while she is pregnant. The young rats forming in her need adequate nutrition and if her diet is lacking in protein, vitamins or minerals the young rats will be deficient to the same degree. Once they are born the damage has already been done, and even if you provide large feedings of the deficient nutrient the rats' bodies won't function as efficiently as it should.

The gestation period (the amount of time it takes the young to develop in the mother) averages twenty-one days, though it can vary. A week or two before the female rat is due to give birth you can feel (gently) a firm swelling in her lower abdomen indicating that she is pregnant. Her nest-building activities will increase in the last days of her pregnancy, so supply her with an ample quantity of nesting material. Clean newsprint, tissue paper (unscented) or an old soft shirt are all satisfactory.

An array of fancy varieties of rats. To avoid indiscriminate breeding, any pair intended for breeding should be kept apart until mating time.

(*Left*) A nest of newborn pink rats. The mother uses the nest to hide her young and to keep them warm. (*Below*) Shown is the genital area of a female (on the left) and that of a male. Note the greater distance between the genital papilla and anus of the male rat.

LACTATION

Once the female rat has had her young she enters the period known as lactation. She produces milk which the young drink while nursing. They attach themselves to the mother's teats instinctively and will need no help from you. The mother rat will do all that is necessary to keep the babies in good health. You must provide her with large amounts of quality food and see to it that drinking water is always available. She has to provide nourishment not only for her own body but also for those of her young.

When the young are first born their mother will spend most of her time

with them, nursing and keeping them warm. You might want to check the young just to make sure they're fine. This does not require handling them and should be done two or three hours after the young are born when the mother takes one of her breaks from nursing them. If she is a pet and knows you, she should not mind. Still, be careful, for she might become defensive and bite your hand. Have someone hold her or distract her attention with a piece of food. Usually the mother will eat any young if they are dead. (Dead young are not usual if the mother has been properly cared for during her

pregnancy.) If the mother rat kills and eats her young it could be that she did not have proper nutrition during pregnancy. If she does this more than once, do not breed from her

aids in keeping them warm and secure.

While the mother is nursing the young, her food intake changes dramatically. She can eat her own weight in food in

The acacia rat of Africa is an arboreal species. It builds a nest on tree tops, not in a burrow in the ground.

anymore. Once you have inspected the young leave them alone until they are ready to venture forth from the nest, usually at the age of two weeks. The mother rat will keep her young carefully hidden in the nesting material. This also

two days. Feed her two or three times what you normally would. Because she is eating more there will be more droppings in the cage. These should be removed. The mother will set up her toilet area away from the nest so there is

A rat when threatened, especially during the breeding period, will not hesitate to assume a standing position. Such an aggressive stance increases a rats stature and possibly frighten a smaller predator.

1

(1) Individual size and weight of the rat at birth depend at least partly on the mother's physical condition and the number of pups in the litter. (2) Like most other mammals, a rat is hairless, blind, and deaf when born. (3) A strong sense of smell guides the still-blind newborn straight to the teats.

2

3

no worry about disturbing the young when you clean the cage. Leave the nest area alone since the mother rat will remove soiled material.

It is extremely important

Hooded rats in the holding cage of a breeder. Understandably, a breeder will keep only those he intends to use for further breedings. Imperfectly marked and poor specimens are removed from his stock.

that you keep the cage clean and sanitary. The young rats are building up immunities during their

first weeks and a clean cage is a great help.

THE YOUNG RATS

Young rats are born pink and hairless. Their eyes are covered by skin and so are their ears. They emit high-pitched bird-like peeps. If there is a large number of young born then they will be smaller in size than those coming from a litter of only five or six. Within a few days a soft coat of fuzz is discernible. In two to four days their ears open and begin to look more like rat ears. Their first teeth appear at the age of eight to ten days. In fourteen to seventeen days their eyes will open. When they are two weeks old they have large heads and squinty eyes and look very much like little puppy dogs. At two-and-one-half weeks they start to explore the nest and follow their mother to see where she goes. They will start to nibble food and play-fight with each other. Make sure the water bottle is within their reach. They require no special food, but you can give them bread soaked in diluted warm milk. Increase the amount of food as necessary to meet the demands of both the mother and her family. While they are exploring the cage you must be careful to remove any sharp object that could cause them to get hurt and remove anything they shouldn't eat that could cause them to get sick.

The young rats are ready to be weaned (leave the mother) at from three-and-one-half weeks to four weeks of age. At this time, if immediate homes are not available, you should keep the males and females separate to prevent any early breedings which might occur.

It is easiest to tell a female rat from a male by checking the litter mates against one another. Generally the distance

between the anus to the genital papilla will be 50% greater in males. At three weeks the distance between the male's penis and anus is one-half inch. The distance between the

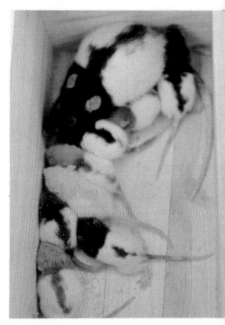

This nest box for rats has solid walls that protect the litter from the cold and draft; it's made of wood, however, certainly not the best choice of materials.

anus and female genital papilla is only three-eighths inch. Also, the male will have a slight swelling where the

1

2

(1) Provide your rat with an object, like a milk carton, to sleep in or hide when frightened. (2) New introductions will upset the "pecking order" in a community cage, but not for long. (3) Keep a close watch when the cage's door is open. You do not want your pet to get lost. (4) Young rats learn from their mother. They simply imitate her actions.

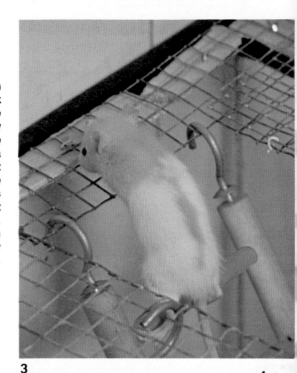

3

4

scrotum will be.

After the young rats have been weaned you should not breed the mother rat for at least one week. She needs time to rebuild her strength and to enjoy her solitude.

ORPHANED RATS

Orphaned rats are not common. If you have orphaned young you have two choices: put them in with another mother rat or raise them yourself. If you have another mother rat with young within one to three days of the same age as the orphans, you can put the orphans in with her young. Put one of the orphaned rats in with the mother and if it appears that she will not accept it, rub her nose with a little Vicks Menthol Vaporub and then put just a little on both batches of young. Do not let the mother nurse too many young rats. If she had eight young of her own and you put another eight in with her, that doubles the number and is too

taxing on her strength. Let her raise a reasonable number and you will have to raise the rest.

You can buy a puppy or kitten milk formula at a pet shop which you can feed to the young. Mix it in a ratio of one to one (with water) if there are no directions. You can feed the formula either from an eyedropper or nursing bottle which many pet shops carry specifically for orphaned animals. It is best to give the young frequent feedings in small amounts. Feed enough until the tummy is white and comfortably full. You will have to use your own judgment to see that you do not overfeed.

You can keep the young warm with a blanket wrapped around a heating pad. (You do not want them to get too warm.) At two-and-a-half weeks you can start feeding them some bread soaked in the formula and cut down on the amount of bottle feedings. Most orphan rats

Coat color inheritance in mammals is a complex affair. Keeping a strain that breeds true is often not possible.

survive quite successfully and they usually make better pets because they have lost any fear of you that might have existed.

SELECTIVE BREEDING

Selective breeding is the process by which man has created much diversity among domesticated animals such as dogs,

horses and cats. Only animals with the desired characteristics are allowed to breed.

When breeding your rats you too have the choice of allowing to mate only the animals that you have carefully selected for the desired qualities. Records can be kept of the sire and doe and their color and

(*Left*) Rats enjoy fruits. Be sure to include some in their diet. (*Below*) This is a sleek, active, and healthy young rat. A sick rat will lose weight and look frail. Isolate a sick rat from other rats you may have.

background. The date and the number and color of the young born from the mating can also be recorded. Though you might view this as a form of work it is very helpful since it is hard to keep every important detail in your head.

Ideas for color breeding rats are many since little work has been done with them. You can try and breed a banded rat by breeding rats with a dark color and white color of the belly which extends up along the rat's side. Continued breeding of such animals could lead to a band across the rat's back.

One could try for a spotted rat by breeding

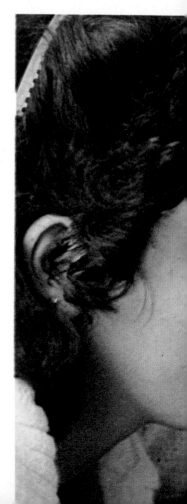

rats with a banded back that has splotches in it. Once the color pattern has become established in the rat's genes try to achieve the pattern in all color phases of the rat.

Many things can be attempted in breeding rats, from inventing new color patterns to different colors becoming dominant in the pet rat hobby. Most important of all, enjoy yourself while breeding your rats. It can be fun and should not be considered as work.

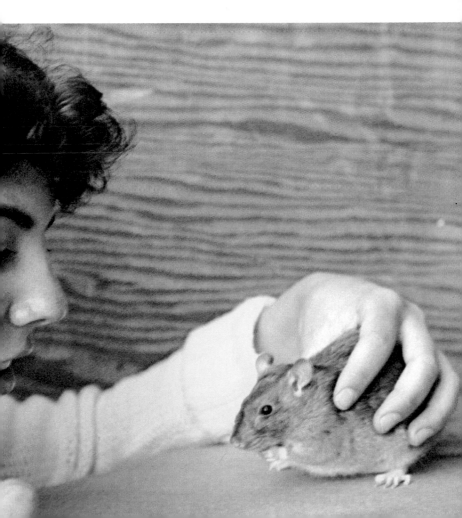

A Healthy Rat

The rat, a strong hardy animal, should live at least two years. A healthy rat's life-span averages two to three years. A three-year-old rat is comparable to a 100-year-old person. It is necessary for his health that you follow the rules of good hygiene. Providing clean bedding and fresh food and water is the most important way to prevent the development of breeding grounds for disease organisms. Almost all rat illnesses develop in an environment of neglect.

COMMON RAT ILLNESSES

The number of diseases and ailments listed below is not meant to show that your animal will get all of them. Instead it shows the common rat illnesses; almost any disease your rat is likely to get is listed. The drugs which can be used in treatment are given but not the exact dosage since that depends on the rat's age, its size and the severity of the case. You should let your veterinarian make the final diagnosis of your pet's illness. Most medicines listed require a presciption. Store all medications in a safe place out of reach of your rat and dispose of them after the expiration date.

Symptoms: Scratching around head and shoulder area causes scabs and bald spots

Disease: Ulcerative dermatitis

Cause: Diet related

Treatment: Include a variety of fruits, vegetables and seeds in the diet

Symptoms: Conjunctivitis (evidenced by red excretions from eye or red circle around eye), chattering, coughing, sneezing, loss of weight

Disease: Chronic murine pneumonia

Cause: Mycoplasma pulmonias (bacteria)

FACING PAGE: A mouse has been placed into a rat's housing quarters— and the rat doesn't like it!

¹ A wheel can be included if there is enough room in a cage. (2) A 20-gallon aquarium is used for this mother rat with her litter. Young rats are active and need room for running and playing. (3) If your rat's coat becomes dirty from lying on soiled bedding, you can clean it with a damp towel or cloth.

2

3

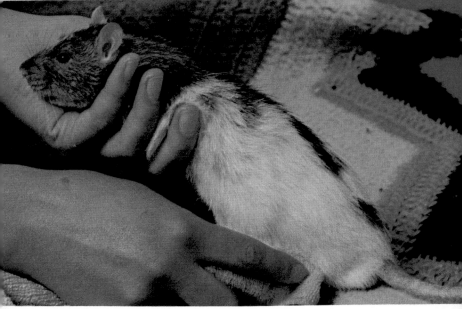

Treatment: Tetracycline licked off the dropper, sulfamerazine in the drinking water or chloramphenicol or Tylosin in prescribed dosage every five days

NOTE: This disease is infectious and difficult to get rid of. Young rats can get infected from their mother. Some rats show a tendency to be immune to the disease while others will get it some months later after being exposed to it. Some rats with this disease can live normally but will eventually die with severe symptoms of the disease.

Symptoms: Red excretions from eye, red circle around eye

Disease: Conjunctivitis

Cause: Infections, irritation, vitamin A deficiency, inflammation from within the eye

Treatment: Antibiotics, Terramycin

Symptoms: Rough coat that won't lie flat, hunched-up back, lethargy, weight loss, conjunctivitis, partial paralysis of hindquarters

Disease: Salmonellosis

Cause: Salmonella (bacteria)

Treatment: Try antibiotics, disinfect surrounding areas

NOTE: If death occurs it is within one to two weeks after the infection. If the rat survives, its growth will be stunted and older rats remain weak and more susceptible to other diseases. Poor hygiene, such as leaving moist foods in cage or not cleaning the cage, is an invitation to this kind of a disease.

Symptoms: Tilting head, circling, difficulty in getting up

Disease: Labyrinthitis

Cause: Bacterial infection of inner ear often associated with upper respiratory infection

Treatment: Antibiotics, steroids

Symptoms: Constrictions with some swelling around feet and tail

Disease: Ringtail

Cause: Humidity below 50%

Treatment: Raise humidity by placing bowl of water near heater outlet; fish tank in room should help

Symptoms: Mammary tumor (in female)

Disease: Neoplasm (growth, usually benign)

Cause: Unknown

Treatment: Surgical removal these tumors usually recur

PARASITES

It is necessary for your veterinarian to take skin scrapings in order to determine whether or not your rat has any of the parasites mentioned below. The skin scrapings are analyzed under a microscope.

Symptoms: Irritated skin, loss of condition

Cause: Lice

Treatment: Use spray meant for birds

NOTE: The life cycle is completed on the host animal. Lice will move to a new animal if the host dies, if they become too numerous or by contact.

Symptoms: Skin lesions, irritated skin, scratching and restlessness

Cause: Mites

Treatment: Use spray meant for birds

NOTE: The female mites burrow under the skin and then lay eggs. The life cycle of the mite is completed on the host

Do not hold your pet too close to your face.

Shown are some color varieties of hooded rats bred today and often sold in pet shops: (1) White and black, (2) cream-colored, and (3) agouti and white.

2

3

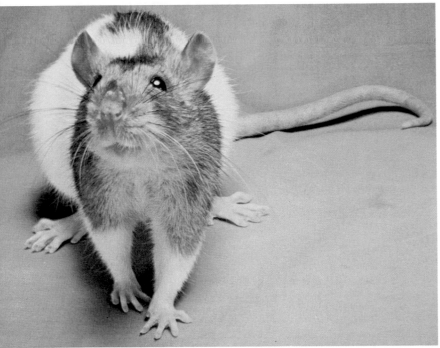

animal. Mites never leave the host except to crawl onto new animals.

Symptoms: Excessive scratching

Cause: Fleas

Treatment: Cat flea and tick powder

NOTE: Fleas can be transferred from other animals in the family that harbor them, such as a dog or cat. The life cycle of the flea takes place on the host animal and its living quarters. The female flea lays its eggs in the animal's den or nest. The larva feed on organic material. It is important to disinfect the animal's living quarters.

Symptoms: Loss of condition, diarrhea

Cause: Pinworms

Treatment: Piperazine citrate

NOTE: The eggs are passed in the animal droppings. Pinworms are infective if eaten. The mature worms live in the cecum and colon. Droppings should be examined to determine if an animal is infected.

Symptoms: Weight loss, diarrhea, pot belly, few signs in less severe cases

Cause: Tapeworm infestation

Treatment: Nichesamide

NOTE: The worms can be contracted through contaminated bedding or from fleas, roaches or beetles.

Symptoms: Dark urine, blood in urine

Cause: Trichosmoides crassicaude (parasite)

Prevention: Good sanitation

NOTE: The adult lives in the bladder or kidneys. The infective stage is carried in the urine. It can result in death.

Diseases can be spread in many ways. Infectious diseases can be transferred through community water bottles, air, contaminated bedding and food, and by your hands. How quickly the illness spreads through a group of rats is determined by the number of rats which have the disease, how infective the carrier rat is, how strong the disease-causing organism is and how resistant the other rats are (whether through natural resistance or resistance acquired through previous contact with the causative organism). Washing your hands after handling the rats will help curtail spreading the disease.

CARE

If your rat does get sick there are certain steps which should be taken to hasten his recovery and to prevent the illness from spreading to any other rats you might have. A sick rat should be isolated immediately from any other rats you have. You can put it in a carrying cage or keep it in his regular cage as long as any obstacles that might bother it are removed, *e.g.,* ladders, bells. The animal should have easy access to both food and water unless your veterinarian directs you otherwise.

(Some illnesses require that both food and water be withheld from your rat.) The food and water should be changed daily and the bedding kept clean and dry. Give the animal a cloth to keep him warm. The rat should be kept quiet and loud noises should be kept to a minimum. The cage in which a sick rat was kept should be sterilized. Soak it in a disinfectant for two to three days.

If you have to treat your pet for any ailment, observe the following precautions. When using a spray either cover the rat's face with your hand to protect his eyes or spray some on a cloth and then apply it to the affected area. When using a powder, care must be taken that neither you nor the rat inhale any of the dust. Hold the cannister down low and again cover your rat's face. Ointments are applied on the infected part of the rat. Always follow any directions you get with the medicine.

(*Left*) An all-wire cage is lightweight and portable. (*Below*) This cage is properly set up for a vacation. Plenty of food and two water bottles situated away from any dry food that could be spoiled if water were to leak on it.

INJURIES

Proper handling can help prevent injuries. Pick up your rat gently without squeezing. If you allow others to hold or play with your rat show them how to properly do so. It is best if you hold your pet and let others watch. Even though your rat is extremely agile and a good climber you must still be careful if you let it climb on any high objects. A fall could cause a broken bone or internal injuries (for example, it could bleed internally and this would go unnoticed by you) and the animal could die.

GROOMING

The rat's skin secretes oils

which it distributes by licking and grooming itself. This gives the coat a glossy shine. Proper nutrition is needed for a healthy skin and coat. Poor appearance of the skin and coat is often a warning that something might be wrong.

Your rat's coat can get dirty from lying on damp shavings. The stain can be removed with a lightly moistened rag. If it is persistent you can buy a spray for cleaning bird's feathers; that should remove the stain.

Your rat's teeth are constantly growing. Each incisor can grow up to five inches each year. However, his teeth get worn down when he eats hard foods. Some old rats' teeth grow too fast and it is necessary to take them

to your veterinarian to have the matter taken care of by a professional.

OLD RATS

Rats go into a decline of old age before they die. Old rats lose weight, their bones become more prominent, and they feel frail to the touch. The fur becomes less dense and fluffier. They tire easily and are content to just sit alongside you and have their back scratched. This can last an indefinite length of time with the rat either dying or the symptoms becoming more severe. If you think your old rat might be suffering, then you might want to have him "put to sleep."

Index

Acacia rat, 68
Accessories (cage), 38–39
African giant rat, 10, 14
African grass rat, 44
Albino rat, 26
Aquarium as cage, 31, 33–34, 82
Bedding, 36–38, 89
Black rat, 11–12
Breeding, 63–64
Breeding age, 64
Breeding room, 63
Brown rat, 11–12
Bubonic plague, 11–12
Bushy-tailed woodrat, 9
Cage (pregnant rat), 65
Cage size, 33

Cages, 33–35
Care, 89
Choosing, 25, 28-29
Chronic murine pneumonia, 80
Cleaning (cage), 40, 42–43
Cleaning (rat), 82, 91
Cleaning (water bottle), 45
Conjuctivitis, 80, 84
Cream-colored rat, 14
Diet (nursing rat), 68
Diet (pregnant rat), 65
Diet (rat), 44–50
Diets (sample), 46–48
Extermination (rat), 17
Facts about rats, 20–21
Fleas, 88

Food, 45–54
Food (amount), 48–50
Gestation, 65
Grooming, 8, 90–92
Handling, 22, 22, 55–56, 56, 85
Healthy rat, 9, 80
Heat cycle, 64
Hooded rat, 6, 15, 28, 72, 86–87
Housing, 33–36
Ilnesses, 80, 84–85
Indoor/outdoor housing, 39–40
Injuries, 90
Introduction, 6
Kangaroo rat, 9–10
Labyrinthitis, 84
Lactation, 66–68
Lice, 85
Long-tailed thicket rats, 50
Mating, 64
Mazes and puzzles, 60–62
Mites, 85
Mole rat, 47
Mycoplasma pulmonias, 80
Neoplasm, 85
Nest box, 73
Newborn rats, 61, 66, 90, 91
Nutritional myctocytic anemia, 52
Old rats, 92
Orphaned rats, 76–77
Packrat, 9
Parasites, 85, 88-89
Pet rat, 23
Pinworms, 88

Pregnancy, 64-65
Rat as pets, 23–25
Rats and man, 16–17, 20
Rats and other animals, 32
Rats (kinds of), 8–12
Rattus norvegicus, 11, 21
Records, 76, 77
Ringtail, 84
Rusty-nosed rat, 21
Salmonella, 84
Salmonellosis, 84
Selective breeding, 77–79
Self-colored rat, 7
Sexing, 66, 73, 76
Skills (rat), 12–13, 16
Snacks, 51–52
Spiny rats, 10–11
Storing food supply 50
Survival (rat), 12–13, 16
Taming, 56–58
Tapeworm, 88
Teeth, 51–52, 91–91
Three-striped rat, 10
Training, 58–60
Transmission (disease), 88–89
Trichosmoides crassicaude, 88
Ulcerative dermatitis, 80
Vitamins, 52–54
Water, 43–45
Water bottle, 35, 43, 73
Wheel, 82
White laboratory rat, 12
Young rats, 72, 73, 76

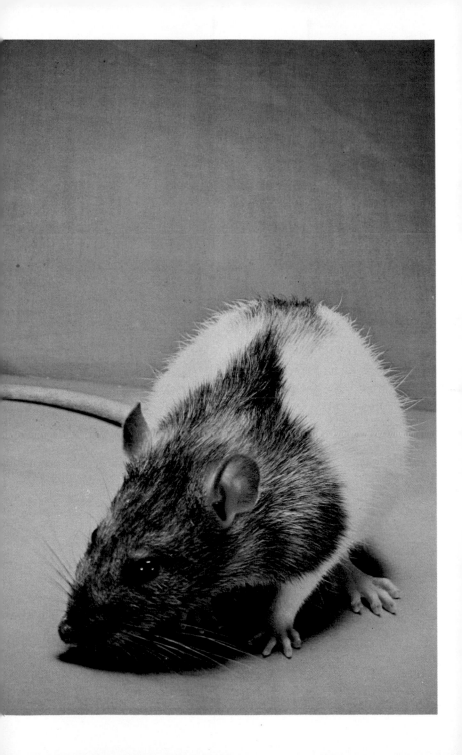

RATS
KW-128